AF510831

INTO Wild Spain

BLACKBIRCH® PRESS

CAPROCK ELEMENTARY LIBRARY

THOMSON GALE™

San Diego • Detroit • New York • San Francisco • Cleveland • New Haven, Conn. • Waterville, Maine • London • Munich

THOMSON
GALE

© 2004 by Blackbirch Press™. Blackbirch Press™ is an imprint of The Gale Group, Inc., a division of Thomson Learning, Inc.

Blackbirch Press™ and Thomson Learning™ are trademarks used herein under license.

For more information, contact
The Gale Group, Inc.
27500 Drake Rd.
Farmington Hills, MI 48331-3535
Or you can visit our Internet site at http://www.gale.com

ALL RIGHTS RESERVED
No part of this work covered by the copyright hereon may be reproduced or used in any form or by any means—graphic, electronic, or mechanical, including photocopying, recording, taping, Web distribution or information storage retrieval systems—without the written permission of the publisher.

Every effort has been made to trace the owners of copyrighted material.

Photo credits: cover, pages all © Discovery Communications, Inc. except for pages 7 © Digital Stock; page 10 © Gallo Images/CORBIS; page 15 © O. Alamany & E. Vicens/CORBIS; page 18 © Ray Tipper/Lonely Planet Images; page 25 © Roger Tidman/CORBIS; page 30–31 © Tony Wharton; Frank Lane Picture Agency/CORBIS; page 38 © Anthony Bannister, Gallo Images/CORBIS; page 43 © Susan Bishop; Papilio/CORBIS

Discovery Communications, Discovery Communications logo, TLC (The Learning Channel), TLC (The Learning Channel) logo, Animal Planet, and the Animal Planet logo are trademarks of Discovery Communications Inc., used under license.

LIBRARY OF CONGRESS CATALOGING-IN-PUBLICATION DATA

Into Wild Spain / Elaine Pascoe, book editor.
p. cm. -- (The Jeff Corwin experience)
Based on an episode from a Discovery Channel program hosted by Jeff Corwin.
Summary: Television personality Jeff Corwin takes the reader on an expedition to Spain to learn about the diverse wildlife found there.
Includes bibliographical references and index.
ISBN 1-4103-0245-8 (hardback : alk. paper) -- ISBN 1-4103-0246-6 (pbk. : alk. paper)
1. Zoology--Spain--Juvenile literature. [1. Zoology--Spain. 2. Spain--Description and travel. 3. Corwin, Jeff.] I. Pascoe, Elaine. II. Corwin, Jeff. III. Series.

Printed in the United States of America
10 9 8 7 6 5 4 3 2 1

Ever since I was a kid, I dreamed about traveling around the world, visiting exotic places, and seeing all kinds of incredible animals. And now, guess what? That's exactly what I get to do!

Yes, I am incredibly lucky. But, you don't have to have your own television show on Animal Planet to go off and explore the natural world around you. I mean, I travel to Madagascar and the Amazon and all kinds of really cool places—but I don't need to go that far to see amazing wildlife up close. In fact, I can find thousands of incredible critters right here, in my own backyard—or in my neighbor's yard (he does get kind of upset when he finds me crawling around in the bushes, though). The point is, no matter where you are, there's fantastic stuff to see in nature. All you have to do is look.

I love snakes, for example. Now, I've come face to face with the world's most venomous vipers—some of the biggest, some of the strongest, and some of the rarest. But I've also found an amazing variety of snakes just traveling around my home state of Massachusetts. And I've taken trips to preserves, and state parks, and national parks—and in each place I've enjoyed unique and exciting plants and animals. So, if I can do it, you can do it, too (except for the hunting venomous snakes part!). So, plan a nature hike with some friends. Organize some projects with your science teacher at school. Ask mom and dad to put a state or a national park on the list of things to do on your next family vacation. Build a bird house. Whatever. But get out there.

As you read through these pages and look at the photos, you'll probably see how jazzed I get when I come face to face with beautiful animals. That's good. I want you to feel that excitement. And I want you to remember that—even if you don't have your own TV show—you can still experience the awesome beauty of nature almost anywhere you go—any day of the week. I only hope that I can help bring that awesome power and beauty a little closer to you. Enjoy!

Best Wishes!

Jeff

INTO Wild Spain

It's definitely the old world, but with some new surprises—wildlife surprises. Join me in a search for creatures that you would never expect to find in Europe.

I'm Jeff Corwin. Welcome to Spain.

Our adventure begins on the Rock of Gibraltar. Despite the fact that it's connected to Spain via the Iberian Peninsula, this land is under British rule and has been so since the eighteenth century. There is an interesting legend about Gibraltar's connection to England. On this great outcrop of rock lives a primate, and it is said that as long as that primate is here, Gibraltar will be a part of England.

As we check this creature out, let's not forget primate protocol, or rules of behavior. We don't want to make any eye contact that tends to startle these animals. I don't want to challenge this animal; I want to observe it. Next to me is the only lower primate living in Europe and the only macaque that lives naturally outside of Asia. This is the Barbary macaque.

These animals are, for the most part, omnivores, which means they eat all kinds of food. They're eating insects and whatever native forbs, fruits, and lichens they can find. They really are beautiful, and it's neat to be this close, but I'm not forgetting that this is a

wild animal that happens to be a male. And male macaques have very big canines. The last thing I want is to be bitten by one of these animals, and that could happen if I didn't respect this animal's space.

These creatures are intelligent, and they use their intelligence just as human beings do to survive in a very tough world. And Gibraltar is tough. There isn't a lot of wild food to be found here. These animals have to work to survive.

Barbary macaques are very intelligent creatures.

Barbary macaques are sociable animals. They depend upon each other for survival, but living together produces tension and stress. What they do to relieve stress is groom one another. Grooming builds bonds between individuals within a group.

These macaques spend 20 percent of the day grooming. To turn your back to an individual and let it touch you is a request for friendship. You need friends to survive on the Rock of Gibraltar.

We've got one more stop in Gibraltar, the city of Gibraltar, the perfect place to launch our great Spanish adventure. From Gibraltar we're going to make our way to Spain

and travel up the Atlantic coast. Eventually we'll arrive at the quaint city of Seville. From Seville we'll head west to one of Europe's most pristine and premier national parks, Doñana. While in Doñana we'll explore its dunes, marshes, and forests.

We're heading about an hour north of Gibraltar to the city of Jerez, which is famous for sherry. Jerez is Spanish for sherry. It is also famous because of the Jerez Zoo, where they are trying to produce one of the rarest felines in the world. We're going to go check those cats out.

This is the Jerez Zoo and Wildlife Rehabilitation Center. What's great about this place is that it's a fantastic introduction to the biology of one of the rarest and most beautiful felines living in Europe.

I've just been sanitizing myself. Why? Because I'm going in this enclosure to show you the crown jewels of Spain here in the Jerez Zoo, and I don't want to contaminate it. This is an animal we might see in the wild, with some luck. But, just in case we don't, I'm going to give you a little flavor of it in the zoo.

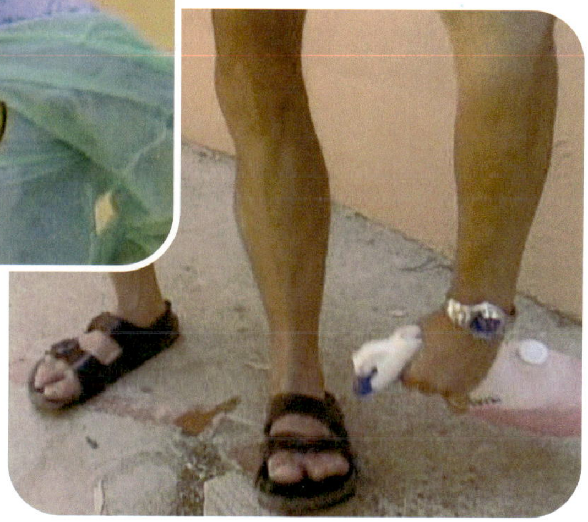

We have come all this way to the Jerez Zoo because of these adorable kittens. These very, very precious animals are Iberian lynx. The smaller one is about eight weeks in age, and the bigger one is about ten weeks old. As with many carnivores, they use play as a way to build up their hunting and survival skills. But what's happening in the zoo is no game. This is a very serious conservation project.

The Iberian lynx just might be the rarest feline on our planet. The goal of the Jerez Zoo is to raise a breeding population of these animals. Unfortunately, there might be less than two hundred wild Iberian Lynx living in Spain. What has happened to these creatures? Why have they been disappearing? As with many animals around the world, the Iberian lynx are facing habitat loss. As the population of human beings in Spain grows, people move into the wild habitat outside of the cities. As a result, these animals lose their homes.

The Iberian lynx is one of
the rarest felines on Earth.

There are other reasons for the disappearance of the Iberian lynx, however. One is a disease that affects their prey. In the 1950s a disease was introduced from Australia into Spain to control the rabbit population. But when the rabbit population dropped, so did the lynx population, because rabbits are the primary prey of the lynx.

The captive breeding at this zoo is just one part of the challenge. Because not only do you need to conserve the cats, but you have to protect their habitat and their prey as well. Now it's time for us to leave these cats and continue our adventure.

We've arrived at Marine Observation Park, right outside the city of Cadiz. This is one of the observation centers. It's a very small place, only eight acres in size, but it's all protected. What's really special about it is that it is habitat to a very unique reptile—an arboreal, or tree dwelling, one at that.

To find one of these creatures, we have to look up into the trees. If we're lucky enough to find one, I think you'll be very surprised. This is one of the types of trees they like to hang out in. It's a eucalyptus, and it's not native to Spain. It smells like Australia. The animals I'm looking for like to hang out in these trees. Unfortunately, I'm not seeing any, and if they're here they're too high up. So let's just go into the interior of this little park.

Wow! A Mediterranean chameleon!

Look at this! That's exactly what I'm looking for, shadowed by the branches and the pine cones. Get down here! This is the creature that I've come to find in this little tract of habitat. This is the Mediterranean chameleon. This one's a good-sized male. There are so many fascinating things about chameleons. For example, they've got stereoscopic vision. Their eyes are on stalks. See that? One eye is looking at you, and the other eye is looking at me! It can look both forward and backward at the same time.

He can look in two directions at the same time.

There's something else that's really cool about them. Check out the feet. See the toes? They're fused in two bundles. These are called zygodactylic feet. There are birds with zygodactylic

He can even look behind himself.

These feet make this chameleon a great climber.

feet, such as macaws, but these are the only lizards with zygodactylic feet. So what is neat about having zygodactylic feet? It gives you an amazing ability to climb.

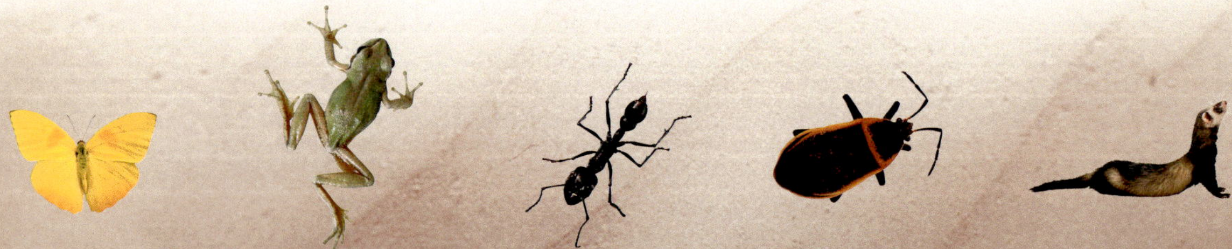

These chameleons also have camouflage. They can change their colors. And there is a lot of research being conducted now to try to find out why and how they change colors. There are many mysteries about these animals. The best mystery of them all is that most people don't know they exist right here in Spain. So we'll release this chameleon and make our way to more adventures.

Only seventy-five miles north is a beautiful and very historic metropolis. This is Seville, one of the most beautiful cities in Europe. It's also one of the oldest cities in Spain. Seville is very historic. And it's also home to one very extraordinary and very secretive bird of prey.

This is the Church of El Salvador. It was built in the 1700s. It's a great place for us to experience some of the history of Seville, and it's also a fantastic place for us to find an amazing animal that lives a secret life in this city. To find that creature, I've got to find a colleague of mine.

The Church of El Salvador

This is Juan Negro. He's an ornithologist, a scientist who studies birds. He'll help us to see another interesting creature that lives among the people of Spain.

We have to go upstairs. Wow, these are some narrow staircases! What's cool about this is that we're surrounded by millions of people, and living among them are birds of prey called kestrels. These are extraordinary creatures. Not only are we going to see them, we're going to experience the research being done to understand and conserve them.

Inside this hole is a fuzzy little chick with a curled beak, a little brown egg, and what appears to be a beautiful adult lesser kestrel, which is a type of falcon. Juan is going to band it. But first I have to capture it with my hand. Juan says the kestrel won't bite, but they do rip apart mice and voles and grasshoppers. I'll go ahead and grab it anyway.

This is a beautiful lesser kestrel! Now here's something cool. Kestrels are famous for their speed when they fly. Look at her head. She can pivot her body and move up and down, while at the same time keeping her head locked in one position like the sights of a gun. This gives her excellent depth perception and field of vision for navigation and striking prey. Falcons are famous for striking prey at tremendous speeds. I'm so excited! I hadn't planned to reach into a hole and pull out a kestrel!

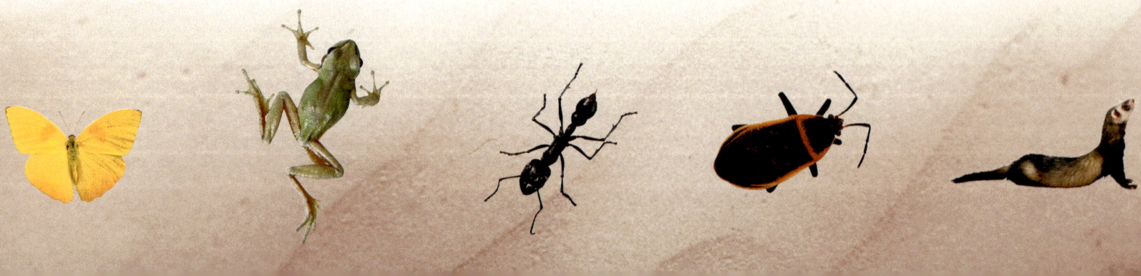

She can move her body while
her head stays locked in place.

We'll band the bird in the shade because the sun is so hot. We'll measure her, too. She's 220 millimeters long. The band on the bird has a unique identification number. By banding this bird, this creature has an identity, and Juan Negro will be able to track it.

He can keep track of its health, its breeding success, and where it goes. Now we'll put it back in the hole. Don't worry little chick, mom's coming! That was awesome! Falcons in Seville. Who would have known?

Now we're on the road toward Doñana. Before we head to that rich biological preserve, we're going to set up a home base in the small village of El Rocio. That's probably sixty kilometers ahead.

Here we have what appears to be an abandoned house. And where there are empty houses, there are sometimes snakes and other creatures. Let's check it out.

It looks pretty promising. Let's just look around and check underneath stuff.

We just might find a very interesting gecko that loves places like this. In fact, it's the largest gecko in Europe. See this? Where there are black scarabs, there are the creatures that eat them, and they are hiding somewhere in the walls of this dilapidated house.

Look at that! That's the animal that lives here. He's taken over this space. It's a Moorish gecko. Now I'll pick it up. Ouch! That doesn't feel nice! I don't like that, Mr. Gecko. Stop biting! He didn't even break the skin. I just saw another gecko that was even larger than this one. But I'm not going to let this one go because it's such a nice find.

Two for the price of one! Look at this. We've got ourselves two Moorish geckos. Ouch! They love to bite. These are really interesting, successful reptiles. You find them living throughout North Africa and across Spain. Geckos

are really cool lizards. They are, I believe, the only group of lizards that have a larynx, or voice box, so they can produce all sorts of sounds. And the famous sound a gecko makes is why it's called what it is. It sounds like this: tat-tat-tat-tat-gecko-gecko-gecko. And I'm not cxaggcrating. It rcally sounds like the word *gecko*. These creatures eat invertebrates such as beetles, larvae, and cockroaches.

Geckos are masters at climbing. They can go straight up a ninety-degree surface, whether it's a rough wall, a stone wall, branches, or even glass. Their fingers fan out and have layers that stick to surfaces.

Geckos can grow brand new tails when their old ones break off.

Then there's the matter of its tail. This lizard does something extraordinary when approached by a predator. When a snake or other predator grabs onto the tip of the gecko's tail, the gecko twists itself and lets the tail break off. The broken-off tail has bundles of nerves, and it twitches about. This confuses the predator. It doesn't know what to make of the situation. It thinks it has killed the gecko, but, in fact, the gecko has outsmarted the predator and escaped. And then, amazingly, the gecko grows a brand new tail.

This just goes to show you that one man's dump is a Moorish gecko's castle. This guy's got all the beetles it wants to eat for the rest of its life right here. So, let's return these amazing animals back to where we found them and continue our journey through Spain.

Now we're back on the road, heading west toward Doñana. And up ahead we're going to stop at our new home base, El Rocio.

This is the small village of El Rocio. And, as you can see, the streets are paved with sand. Why?

Well, look how I'm getting to Doñana—on horseback. Take one horse, three snakes, one pinch of good-looking turtle, combine, shake well, and what do you get? Doñana.

El Rocio...where the streets are paved with sand.

Whoa. Well, here we are. Welcome to Doñana National Park. This is easily one of Europe's most prestigious and certainly one of the most biodiverse national parks you'll find in this part of the world. Why? Because this place brings together a great diversity of habitats. For example, it has forest, grasslands, dunelike deserts, and coastlines.

The water isn't very deep here, but it's very murky and muddy.

Here's a European pond turtle. It's a big, beautiful, dramatic turtle. And look at those spots! See those bright eyes? I love the eyes of these turtles. This one is a female. I think this is a real pretty turtle. It has those nice yellow spots on its body and great pupils.

And of course, there's that awesome turtle smile.

There is something else over there, so let's put him down and check it out.

Aha! Here are three beautiful snakes. These are European water snakes, or pond snakes. My guess is that we interrupted some sort of courtship behavior. In many species of snakes, males are smaller than females. Here, we might have two males pursuing one female. Females produce pheromones, which are chemicals whose smell attracts male snakes.

If you spend a lot of your life in the water, and you're moving though the murkiness, you want to be able to see, right? Check out those eyes. They're positioned at the top of this animal's head. So as it's careening through the water, its body can be submerged and it can still see. You also would want to find food. These snakes eat frogs and crayfish, as well as young birds and rodents that venture too close to the water's edge. They also eat large water beetles and the larvae of amphibians. These snakes need to conceal their bodies, and they do it with excellent camouflage. It's a perfect match for this sludgy algae growing around them.

What I really like about these snakes is that they're very mild tempered. Back home when I was a kid and I would find a banded water snake, I'd come home with this thing hanging off

European pond snakes are pretty mellow.

my arm. My mom would panic, and then I'd let it go. These animals are making no attempt to bite me. I don't know if that's because they're still a little chilled, or if they're just really gentle. These are wonderful snakes and a great example of the secrets you find in Spain. These animals live in ponds throughout this country as well as the rest of Europe. When people are sitting on their blankets near the water, they don't even know that these creatures are swimming right next to where they sit. Let's set them free and continue moving onward.

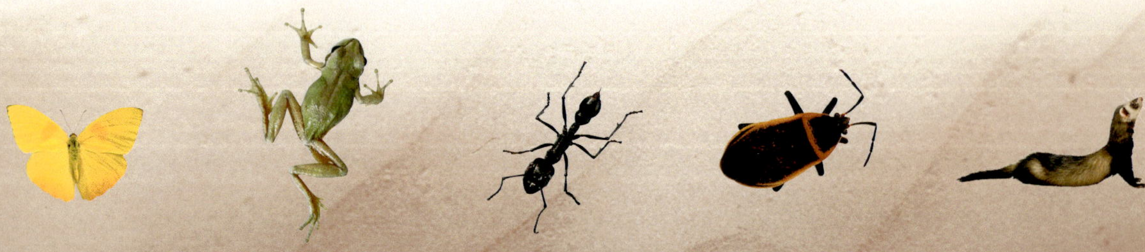

Look at this absolutely beautiful terrain. This is another example of the unexpected here in Spain. When we think of the landscape of this region, what usually comes to mind are grasslands, pastures, and windmills. But look at this beautiful dune system. What's great about exploring Doñana is that it's got four major ecosystems, and this is one of them—the dunes. They sweep their way to the ocean. It's excellent habitat for wildlife. To track the animals we're looking for, we look for movement in the sand.

There are deer tracks here. Sand has blown over them, but you can still see where the animals have dragged their hooves through the sand as they moved. See this right here? This is from *zorro*, a fox.

Check this guy out! This is a very interesting creature, the sacred scarab, or infamous giant dung beetle. What it's rolling is not a mud ball, but a dung ball. It's probably from one of the wild horses that live here in Doñana. What the beetle has done is shape this massive pile of dung into an almost perfect sphere. It's pushing the ball with its back legs. It'll push this giant ball of dung, which is five or six times bigger than itself, for a great distance. Maybe it's not a great distance for us, but for a creature this small, it's a lot of work. Instinct drives this creature to tumble and roll towards its ultimate destination, which will probably be over this bluff. See what it's doing now? Look at this; it's excavating a good area for that dung.

What the beetle will do is take the dung to the other side of this bluff, and then push it down into a tunnel it has dug. It will then deposit an egg in this ball of dung, which will be the primary source of food for the larva. What's really cool about this is that this dung beetle shows how organisms are connected. Nothing is wasted. What began as dung for a large creature becomes a valuable resource to another

It's plowing through the sand with its head.

organism, in this case a very beautiful but somewhat stinky dung beetle. Look at the front of the head of this creature. It's like a bulldozer. It uses its head to plow through the sand to excavate and make a path for this big pile of dung.

The front of its head is just like a bulldozer.

There's something else that's very interesting about this sacred scarab, this dung beetle. It's not only transporting dung, it's also transporting other creatures. Because as it moves these great distances, this insect becomes a bus for parasites. If you

look at its belly, you'll notice little flylike creatures on the abdominal surface. The parasites appear to have wings, but they don't need them. Why fly if you can get a free ride? And in the case of these little bugs, their transportation is this beetle. Let's leave this creature to its dung and continue our trek.

Check out the flylike creatures on its belly.

Our job is tough. We're in the field all the time and are traveling everywhere. So we need a break. And this is the best way to do it—experience a bit of Spanish culture. We get the food, and of course, the ancient art of flamenco. More, please! Más, por favor!

Time to take in some Spanish food and culture.

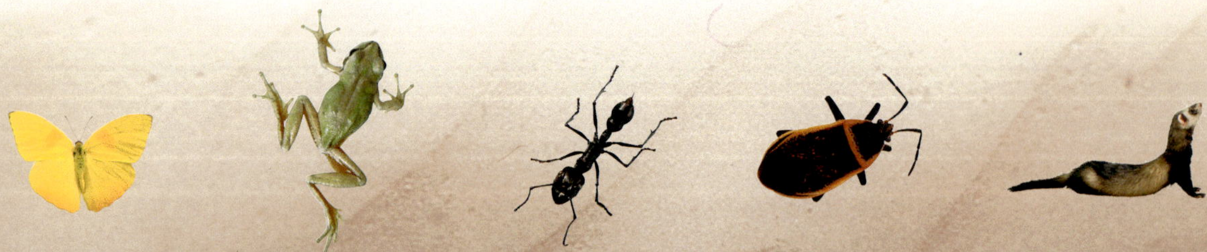

Wow! We move directly from the dunes to the forest. These dunes are literally moving inland from the ocean and swallowing up the habitat. But this is where we want to go. This is a Mediterranean pine forest.

We have a cavity right here, and we have definite evidence that a creature entered it. It's probably the creature that excavated it. If I were in Asia or Africa, the first thing that would come to my mind is a monitor lizard, *Veranidae*. But this was not created by a verani because they don't live in Europe. The size of this hole, as well as the nature of the foot and tail prints and the belly that was dragged through here, tell me that it can be only one kind of lizard. It's a lizard with powerful jaws, so I don't want to be bitten. I'm just going to carefully feel in there.

This hole is definitely occupied. I'm going to reach in there and grab it behind the head. Don't bite me! All right, I've got it! Now he's putting a little resistance up. Let go of those roots! All right, there we go. Check this lizard out. This is the extraordinary oscillated lizard. Now it's assuming a defensive posture with its tail twitching and its mouth open, showing you that bright pink mouth.

Look at that bright pink mouth!

Why is it called the oscillated lizard? It's because of these beautiful bright blue spots. If you look at the thighs, you can see its spurs, which tells us that this creature is definitely a male.

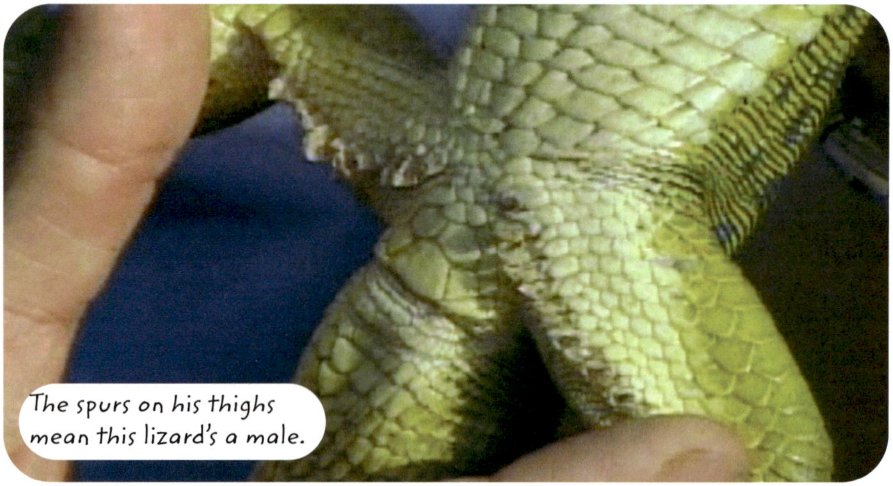

The spurs on his thighs mean this lizard's a male.

It's huge! This is a two-and-a-half-foot lizard, and its body is bigger than its tail. And if you look at the surface of the tongue, you'll see that it's slightly forked, like a serpent's. The oscillated lizard is easily Europe's most dramatic and certainly largest lizard.

Now, this looks really promising. I just moved this log and something made a sound. I hope it's not a hornet's nest! We've definitely had some activity here. This hole might have been made by a European boar looking for grubs. But that's not what I'm interested in.

Something under here made a sound. There's a head sticking out. This is definitely a venomous viper.

Wow! Did you see that? Look at that perfect camouflage. Now he's on the move. This beautiful serpent is Lataste's viper. They often say that European vipers are slow, but this guy is moving. He's been energized by the sun.

My hands are shaking! Look how beautiful that snake is. It's got that amazing rostrum, or snout, that curls up like a rhino's. Little projections called keels run lengthwise against the back of each scale. And the belly is almost cinnamon-like. This animal eats many of the creatures we've discovered on this adventure, such as the Moorish gecko we discovered in that old abandoned house. This creature loves to eat geckos—not when they're up in the trees, but when they come down to earth. This is a wonderful discovery! Now is probably a good time for us to release this animal back where we captured it.

This is magical stuff! Well, that's it folks. We traveled through Spain, and we discovered some of its most amazing creatures. So until we meet again, I'll be very much looking forward to our next adventure together.

Glossary

arboreal an animal that spends most of its time in trees

biodiverse having a variety of life forms

carnivores meat eaters

conservation preservation or protection

ecosystem a community of organisms

habitat a place where animals and plants live naturally together

invertebrates animals without a spinal column

pheromones chemicals released by female animals to attract males for breeding

primate a type of mammal such as monkeys, apes, or humans

pristine clean and unspoiled

viper snake

Index

Arboreal, 17

Beetle, giant dung, 38, 45
  parasites and, 39–40

Camouflage, 21, 44
Chameleon, Mediterranean
  camouflage, 21
  eyes, 19
  feet, 20
Conservation, 14, 16

Gecko, Moorish, 28, 45
  sounds, 29
  tail, 31
Gibraltar, Rock of, 6, 11

Habitat, 14, 16, 17, 19, 33, 37, 41

Kestrel, 23, 24
  tracking, 26

Lizard, oscillated, 42–43
Lynx, Iberian, 14

Macaque, Barbary
  food of, 8
  groooming, 11
  teeth, 9

Predator, 31
Prey, 16, 22, 24

Rabbit, 16

Scarab
  black, 27
  sacred. *See* beetle, giant dung
Snake, European water, 34–36

Turtle, European pond, 34

Viper, Lataste's, 44–45

***Z**orro* (fox), 37

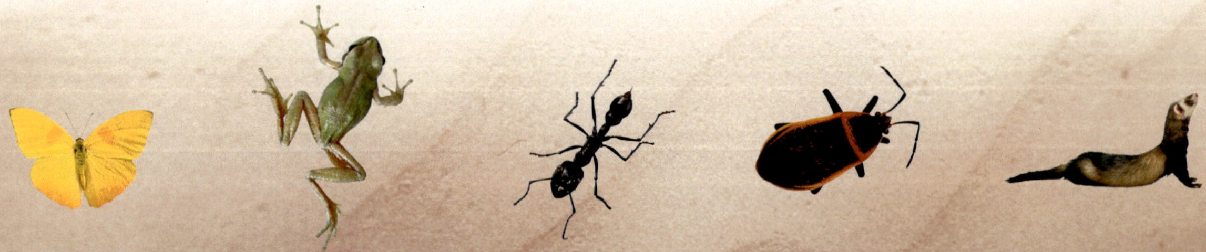

CAPROCK ELEMENTARY LIBRARY